Disclaimer

The publisher of this book is by no way associated with the National Institute of Standards and Technology (NIST). The NIST did not publish this book. It was published by 50 page publications under the public domain license.

50 Page Publications.

Book Title: Advanced Mass Calibration and Measurement Assurance Program for State Calibration Laboratories NISTIR 5672 2005 Ed

Book Author: Georgia L. Harris; Kenneth L. Fraley

Book Abstract: This publication provides guidelines for evaluating data from advanced mass calibrations and for establishing measurement assurance programs in precision mass calibration laboratories. The NIST Weights and Measures Division (WMD) will use these guidelines when evaluating advanced mass calibration data for State laboratories that request technical support, Recognition, and/or National Voluntary Laboratory Accreditation Program (NVLAP) accreditation. Advanced mass calibrations use weighing designs, such as those found in NBS Handbook 145 (SOP 4, 5), NISTIR 6969, Selected Publications, NBS Technical Note 952, and the NIST/SEMATECH e-Handbook of Statistical Methods that require the use of computer software (mass code) for the data reduction. These weighing designs are normally used when high precision (low uncertainty) mass measurement results are sought, although weighing designs can be used at any uncertainty level. The uncertainty reported using advanced weighing designs is based on the historically observed process of similar measurements and is very dependent upon correct procedures for defining these processes.

Citation: NIST Interagency/Internal Report (NISTIR) - 5672

Keyword: advanced mass;mass calibration;measurement assurance;state laboratories;weighing designs

NISTIR 5672

Advanced Mass Calibration and Measurement Assurance Program for State Calibration Laboratories (2005 Ed)

Fraley, Ken L.
Harris, Georgia L.

National Institute of Standards and Technology
Technology Administration, U.S. Department of Commerce

NISTIR 5672

Advanced Mass Calibration and Measurement Assurance Program for State Calibration Laboratories (2005 Ed)

Fraley, Ken L.
Oklahoma Bureau of Standards

Harris, Georgia L.
*Weights and Measures Division
Technology Services*

March 2005

U.S. DEPARTMENT OF COMMERCE
Carlos M. Gutierrez, Secretary
TECHNOLOGY ADMINISTRATION
Phillip J. Bond, Under Secretary of Commerce for Technology
NATIONAL INSTITUTE OF STANDARDS AND TECHNOLOGY
Hratch G. Semerjian, Acting Director

March 16, 2005

Preface

This publication was originally written by Ken Fraley, metrologist with the State of Oklahoma, and Georgia Harris, physical scientist with the NIST Weights and Measures Division. Ideas from final users regarding publication content were sought at the 1994 National Conference on Weights and Measures meeting held in San Diego, CA. The publication was written to provide guidance for calibration laboratories in their desire to provide improved precision mass calibrations and for the NIST Weights and Measures Division to ensure uniform evaluation of laboratories seeking to make mass measurements and be accredited at the advanced level of mass calibration.

Since the first edition in 1995, numerous practical questions have been raised and additional input has been sought from mass calibration experts. This edition seeks to enhance the original publication and provide additional guidance. Copies of Standard Operating Procedures 5 and 28 (3-1 Weighing Designs and Advanced Weighing Designs respectively) are included as an appendix. Additional weighing designs, equations for between-time standard deviations, and updates for uncertainty analysis are included as well.

About the Authors

Ken Fraley is a metrologist with the State of Oklahoma. He has over 10 years of experience in making precision mass measurements using weighing designs as described in this publication. He has carried out extensive experimentation and implemented measurement assurance programs to ensure that measurements made in the Oklahoma laboratory are consistent and uniform with those at the national level. As background experience, Ken developed the first draft of this document based on discussions with NIST and early users of weighing designs in the State laboratories, and provided a critical review of Advanced Laboratory Auditing Program (LAP) problems submitted after the first Advanced Mass Metrology seminar in 1993. He has also coordinated and analyzed several interlaboratory comparisons among laboratories working at the advanced level described in this publication. He has analyzed data, prepared preliminary and final analyses, and presented results of many interlaboratory comparisons conducted at basic, intermediate and advanced levels. In this edition, Ken has clarified ideas that now have had over 10 years of refinement, has introduced extensive spreadsheet usage, and provided additional graphic content.

Georgia Harris is the Group Leader of the Laboratory Metrology Group in the NIST Weights and Measures Division. She provided direction and encouragement to Ken in developing the ideas and provided editorial support for the initial publication. In this revision, she has provided updated copies of SOP 5, SOP 28, ideas and content for spreadsheet analysis, and enhancements based on answering many technical questions from laboratories working at this level, and from reviewing numerous annual submissions for laboratories seeking formal Recognition at this level. The Weights and Measures Division is responsible for providing technical support and guidance to the State legal metrology laboratories to ensure uniformity in the legal metrology measurement infrastructure; this publication is intended to provide support and guidance not only for State weights

March 16, 2005

and measures laboratories, but also for other calibration laboratories seeking to implement advanced weighing designs.

The authors wish to express their thanks to M. Carroll Croarkin (NIST) for providing between-time standard deviation formulae and assistance regarding updating mass calibration uncertainties to meet the ISO Guide to the Expression of Uncertainty in Measurement, to Jerry L. Everhart (JTI Systems, formerly with EG&G Mound) for providing guidance in Process Measurement Assurance Programs, and to all of the metrologists who regularly participate in WMD training and regional meetings for their questions, comments, and desire to make precision mass measurements to the best of their capabilities. For the 2005 update, the authors greatly appreciate the efforts of Hung-kung Liu of the NIST Statistical Engineering Division in supplying missing factors for the between-time standard deviation equations and for his critical review of the document.

March 16, 2005

Contents

Preface iii
Program Objective 7
Program Prerequisites 8
    Training 8
    Facilities 8
    Equipment 9
    Standards 10
Advanced LAP Problems 11
    LAP Problem 1 12
    LAP Problem 2 12
    LAP Problem 3 12
    Follow up 12
Establishing Measurement Controls 15
    Process Evaluation 15
    Data Input 17
    Handling the Output 18
Reviewing Mass Code Report 19
Graphs and Control Charts 20
    Critical Graphs 20
    Optional Graphs 21
Proficiency Tests 21
Evaluation Criteria for Proficiency Tests 22
    Verification of Laboratory Values 23
    Verification of the Laboratory Precision 23
File Management 23
Software Management 24
    Distribution 24
    Licensing and Software Quality Assurance 24
    Updating 24
    Approved Weighing Designs 24
Documentation of Standard Operating Procedures 25
Traceability and Calibration Intervals 25
Formulae and Calculations 26

March 16, 2005

March 16, 2005

Advanced Mass Calibration and Measurement Assurance Program for State Calibration Laboratories

Program Objective

This publication provides guidelines for evaluating data from advanced mass calibrations and for establishing measurement assurance programs in precision mass calibration laboratories. The NIST Weights and Measures Division (WMD) will use these guidelines when evaluating advanced mass calibration data for State laboratories that request technical support, Recognition, and/or NVLAP accreditation.

Advanced mass calibrations use weighing designs, such as those found in NBS Handbook 145 [1] (SOP 4, 5), NISTIR 6969, Selected Publications [2], NBS Technical Note 952 [3], and the NIST/SEMATECH e-Handbook of Statistical Methods [4] that require the use of computer software (mass code) for the data reduction. These weighing designs are normally used when high precision (low uncertainty) mass measurement results are sought, although weighing designs can be used at any uncertainty level. The uncertainty reported using advanced weighing designs is based on the historically observed process of similar measurements and is very dependent upon correct procedures for defining these processes.

NIST calibrations provide traceable standards at one point in time. The major advantage in this program is the ability to evaluate reference/working standards and the measurement process over time, providing ongoing assurance regarding accuracy and traceability of the mass standards for both the laboratory and its customers. Ongoing evaluation of the measurement process provides the laboratory with data that can be used to establish or adjust calibration intervals for reference/working standards. The measurement assurance program is also critical for defining and reporting realistic uncertainties.

---

[1] Taylor, John K. And Henry V. Oppermann, NIST [NBS] Handbook 145, Handbook for the Quality Assurance of Metrological Measurements, November 1986.

[2] Selected Laboratory and Measurement Practices, and Procedures, to Support Basic Mass Calibrations

[3] Cameron, J. M., M. C. Croarkin, and R. C. Raybold, Technical Note 952, Designs for the Calibration of Standards of Mass, June 1977.

[4] *NIST/SEMATECH e-Handbook of Statistical Methods*, http://www.itl.nist.gov/div898/handbook/, 2005.

March 16, 2005
Program Prerequisites

The following items are listed as general guidelines for a laboratory conducting an internal evaluation of its program for suitability in the advanced mass calibration program for States. These guidelines have been established based on good measurement practices, good laboratory practices, and a similar fee-funded program (NIST Mass MAP) operated by the NIST Mass and Force Group. Many specific technical recommendations are taken from NIST Handbook 143, State Weights and Measures Laboratory Program Handbook, G. Harris, Editor, March 2003, and NIST Handbook 150-2G: NVLAP Calibration Laboratories, Technical Guide for Mechanical Measurements, C. Douglas Faison and Carroll S. Brickenkamp, Editors, March 2004. Deviations from recommendations are occasionally made when data is available to support it; however, judgments should be made carefully when evaluating data, since some deviations from these practices will inadvertently increase measurement uncertainties and may contribute to measurement errors.

Training

< Satisfactory completion of an Intermediate Metrology Training course and Laboratory Auditing Program (LAP) problems within the last five years are expected before attendance at the Advanced Mass Training course which is traditionally taught in odd-numbered years.

< Although required for NIST Recognition, attendance at the regional measurement assurance program meetings is not required to perform advanced mass calibrations. However, regular updates on precision mass procedures and issues are often provided at the regional meetings.

< Satisfactory completion of the Advanced Mass Training course and Advanced LAP problems and successful application of these guidelines are expected before Recognition or accreditation at the Echelon I level as described in NIST Handbooks 143 and 150-2G.

Facilities

< Environment -- The temperature for the laboratory where mass measurements are made should be selected at a point between 20 $^{\circ}$C and 23 $^{\circ}$C, with allowable variation of $\pm$ 1 $^{\circ}$C (e.g., 22 $^{\circ}$C $\pm$ 1 $^{\circ}$C), with a maximum change of 0.5 $^{\circ}$C per hour. Air flow must be low enough so that it does not interfere with balance or mass comparator operation. Humidity should be set between 40 % and 60 % relative humidity (e.g., 45 % $\pm$ 5 %). Environmental conditions must be monitored according to technical criteria and measurements should not be made when prescribed conditions are not maintained. Deviation from stated environmental parameters requires a thorough evaluation of the impact of the deviation on measurement errors and uncertainties.

< Vibration -- The laboratory location and design should be such as to avoid or minimize potential sources of vibration that will interfere with precision mass calibration.

March 16, 2005

< Cleanliness -- Good housekeeping practices and cleanliness specifications are found in NIST Handbook 145, 143 and NIST/NVLAP Handbook 150-2G. Contamination from dust, hair, paper, shipping/packaging/storage materials like felt and velvet, and other air contaminants has been found to be a critical concern.

Equipment

< Computer and printer -- A computer of sufficient memory and processing ability is essential. In the laboratory, every effort must be made to minimize the impact of introducing temperature gradients in the measuring areas. Laptops or other wireless or panel monitors are preferred to large heat-producing systems. The printer should be capable of routinely producing the 40-page reports generated by the mass code, and it should also be capable of printing graphics. The printer should not be located in the precision mass laboratory.

< Balances -- A list of the laboratory's balances and process control chart data should be submitted to WMD for evaluation. Control data for each balance should consist of the following:

- Balance manufacturer, model number and serial number;
- Capacity and resolution; and
- Pooled standard deviations (accepted within-process standard deviations, s_w, and accepted long term standard deviation of the check standard, ϑ, showing number of degrees of freedom, loads, and specific weighing designs.

Laboratory balances will be evaluated to determine their suitability for this program. Minimum balance performance specifications recommended for this program are as follows:

Loads	*Standard Deviation should be ≤:
20 kg	1.5 mg
10 kg	0.20 mg
1 kg	0.050 mg
600/500/400/200 g	0.025 mg
60/50/40/20 g	0.0050 mg
6/5/4/2 g	0.0025 mg
600/500/400/200 mg	0.0013 mg
60/50/40/20 mg	0.00050 mg
6/5/4/2 mg	0.00030 mg

* The standard deviations in this table were developed by calculating 1/3 of OIML Class E_1 tolerances, accounting for typical uncertainties associated with the standard and other factors, and developed by working backwards from a target expanded uncertainty. Many laboratories will not have balances or processes that can achieve these results on a routine basis.

March 16, 2005

- Barometer -- A barometer having documented accuracy of $\pm$ 65 Pa (0.5 mm Hg) with evidence of traceable measurement results (typically from an accredited laboratory) must be available.

- Thermometers -- Thermometers to measure air temperature having accuracy of $\pm 0.1\ ^{\circ}$C with evidence of traceable measurement results (typically from an accredited laboratory) are required. Temperature measurements made at this level are generally made within the balance chambers.

- Hygrometer/Psychrometer -- Percent relative humidity should be measured with an accuracy of $\pm$ 5 % and have evidence of traceable measurement results (typically from an accredited laboratory).

- An environmental recording device is critical for monitoring laboratory conditions. Even though a number of environmental corrections are made in the mass measurement process, ensuring environmental stability during the 24-hour period preceding a calibration (particularly for temperature) is important to ensure proper thermal and environmental equilibration of mass standards.

Standards

- Reference (formerly called Primary) Standards -- A minimum of two 1-kg reference standards (four 1-kg reference standards are recommended) with NIST calibration and density determination are needed. Calibration values should be less than two years old. Standards should be calibrated at least every two years unless a measurement assurance program that monitors the reference kilogram standards is in place and demonstrates ongoing stability and validity of the mass values. If only two reference standards are available, and if they are used with equal frequency, the measurement assurance program will not be considered adequate without some type of verification using standards from outside the laboratory that have recent NIST-calibration. GMP 11, Good Measurement Practice for Assignment and Adjustment of Calibration Intervals for Laboratory Standards, and GMP 13, Good Measurement Practice for Ensuring Traceability (or other equivalent procedure), should be implemented in the laboratory to document the traceability hierarchy and calibration intervals that the laboratory will follow. (See discussions on Proficiency Tests and Graphs and Control Charts.)

- Check standards -- Check standards (sometimes called control standards) are not required to be calibrated by NIST. However, having the check standards calibrated by NIST or a competent external source provides an effective mechanism to identify and evaluate biases that may be occurring in the measurement processes that would otherwise go undetected; having an external calibration is essential.

March 16, 2005

> The check standards must be one-piece design (to provide the necessary stability and act as surrogates to the reference standards) with known densities and have been assessed to comply with limits on magnetic susceptibility as required in ASTM and OIML standards, in the following decade denominations: 1 kg, 100 g, 10 g, 1 g, 100 mg, 10 mg, 1 mg. "ASTM Class 1, Type 1, Grade S" or "OIML Class E_2" verbiage may be used if specifying weights for purchase. Additional check standards above 1 kilogram are needed to handle the entire range of calibration services, e.g., 10 kg.

Measurement Assurance (Control)

< The laboratory needs to have a measurement assurance (control) system already in place before trying to perform advanced mass calibrations. Practical, hands-on experience in the laboratory is essential to making good mass measurements. A current measurement assurance system and data are essential for demonstrating measurement proficiency and/or justifying any deviations from these recommendations.

Advanced LAP Problems

See Figure 1 for a graphic view of the components required in a complete analysis of the Advanced LAP problems.

Laboratory Auditing Program (LAP) problems are used for establishing a baseline for the initialization of the check standards, to provide initial data to assess the between-time component of the measurement process, for providing validation on uncertainty statements, and for evaluating the proficiency level at which the laboratory uses the mass code. The data collected in the LAP problems is reduced by each laboratory using the mass code. Each qualified metrologist must complete training and the Advanced LAP problems and be able to reduce and analyze their own data. Each laboratory is responsible for graphing and analyzing data when determining the "in control" or "out of control" condition of their standards (see sections on Establishing Measurement Control and Graphs and Control Charts). Observed surveillance values must be compared to the reported NIST values to determine the level of control. A copy of all data, data files, reports, graphs, and final analysis are to be sent to WMD for evaluation along with the most recent NIST calibration report for the standards used. The final written analysis, demonstrating a thorough understanding of the measurement assurance system and uncertainty analysis at this level of work will be considered as the most critical component of the completed Advanced LAP problems.

NOTE: Standards should not be cleaned using solvents during the initial data-collection period as these tests will provide data for determining the total and between-time standard deviations for each series. Cleaning standards changes their mass values and may invalidate the calibration. Cleaning plans and procedures must be documented as a part of the laboratory procedures. At least 7 to 10 days are required for environmental equilibration on standards that have been cleaned with solvents prior to recalibration.

March 16, 2005

LAP Problem 1: Ten (10) complete runs on reference/working State standards from 1 kg to 1 mg. (Initial data from the first one or two runs may be submitted to be sure the laboratory is on the right track.)

LAP Problem 2: Two (2) complete runs on standards from 30 kg to 2 kg.

The series selected when ascending from 1-kg may be the same as those used when descending; however, a single restraint is usually used and the between-time standard deviation formulae must be derived.

If the laboratory maintains both metric and avoirdupois standards with NIST traceability, one additional LAP problem should be conducted. The avoirdupois standards start at 1 pound and usually will require a "cross over" from the kilogram. Special tare weights (92.815 g) can be obtained to facilitate this process.

LAP Problem 3: Two (2) complete runs on standards from 50 lb to 1 μlb.

Follow up: Note measurement assurance guidelines and traceability for using the initial LAP problems in continuing measurement assurance.

Note: Two complete runs is not adequate to provide data for establishing initial limits nor for establishing a baseline for acceptable measurement assurance nor for validating uncertainties at this level. If the laboratory plans to provide internal calibration results or extended service to customers, additional data must be obtained and analyzed. Process statistics determined with limited data must have uncertainties that reflect the actual degrees of freedom. The laboratory may limit their application of this process to the smaller mass levels (e.g., 1 kg to 1 mg, or 100 g to 1 mg). LAP problems 1 and 2 may be combined if a limited range will be used for this level in the laboratory. For example, ten runs from 10 kg to 1 mg, would be another acceptable approach. If the laboratory does not plan to use the mass code for avoirdupois standards, LAP problem 3 is not required.

March 16, 2005

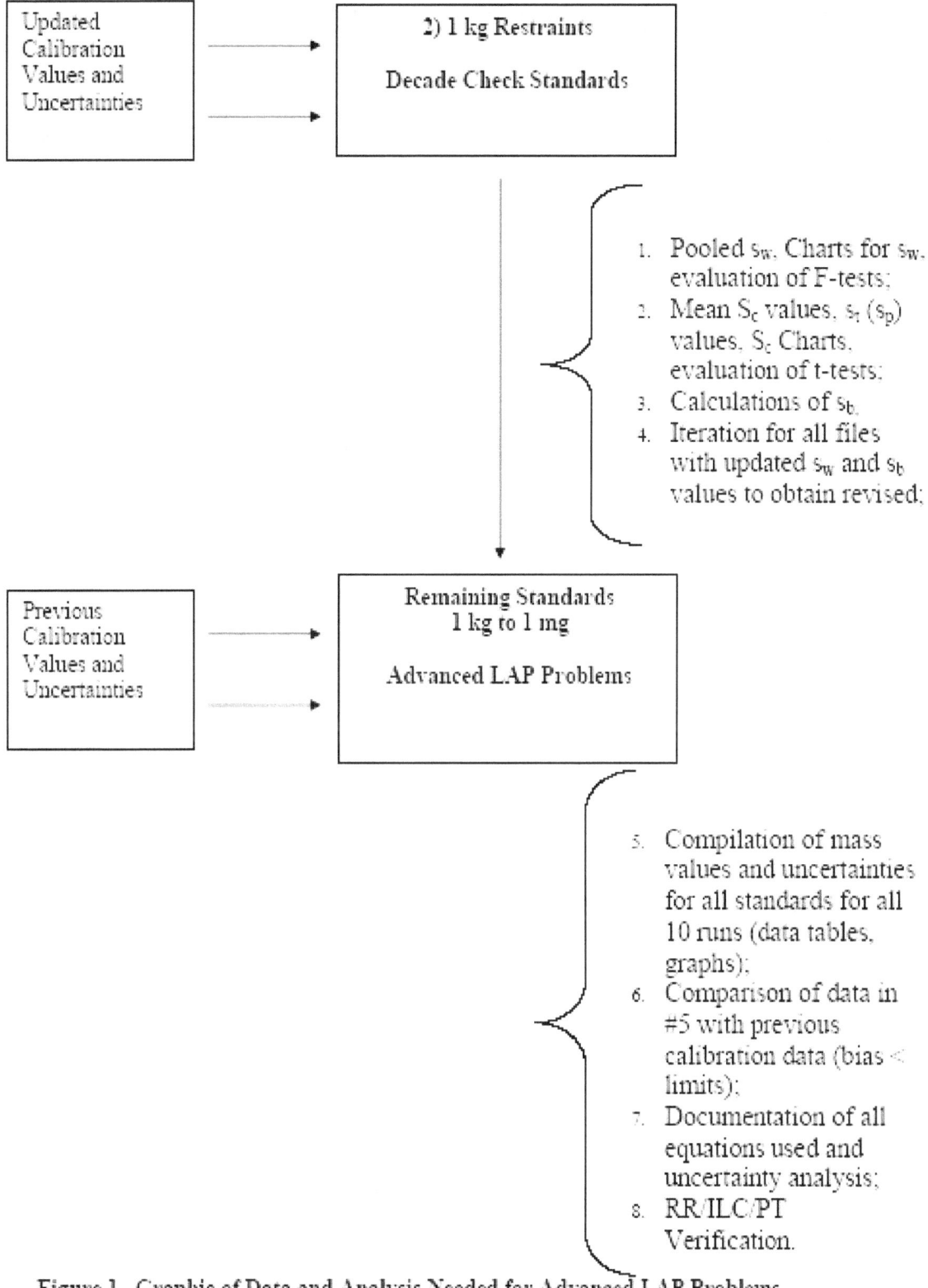

Figure 1. Graphic of Data and Analysis Needed for Advanced LAP Problems.

March 16. 2005

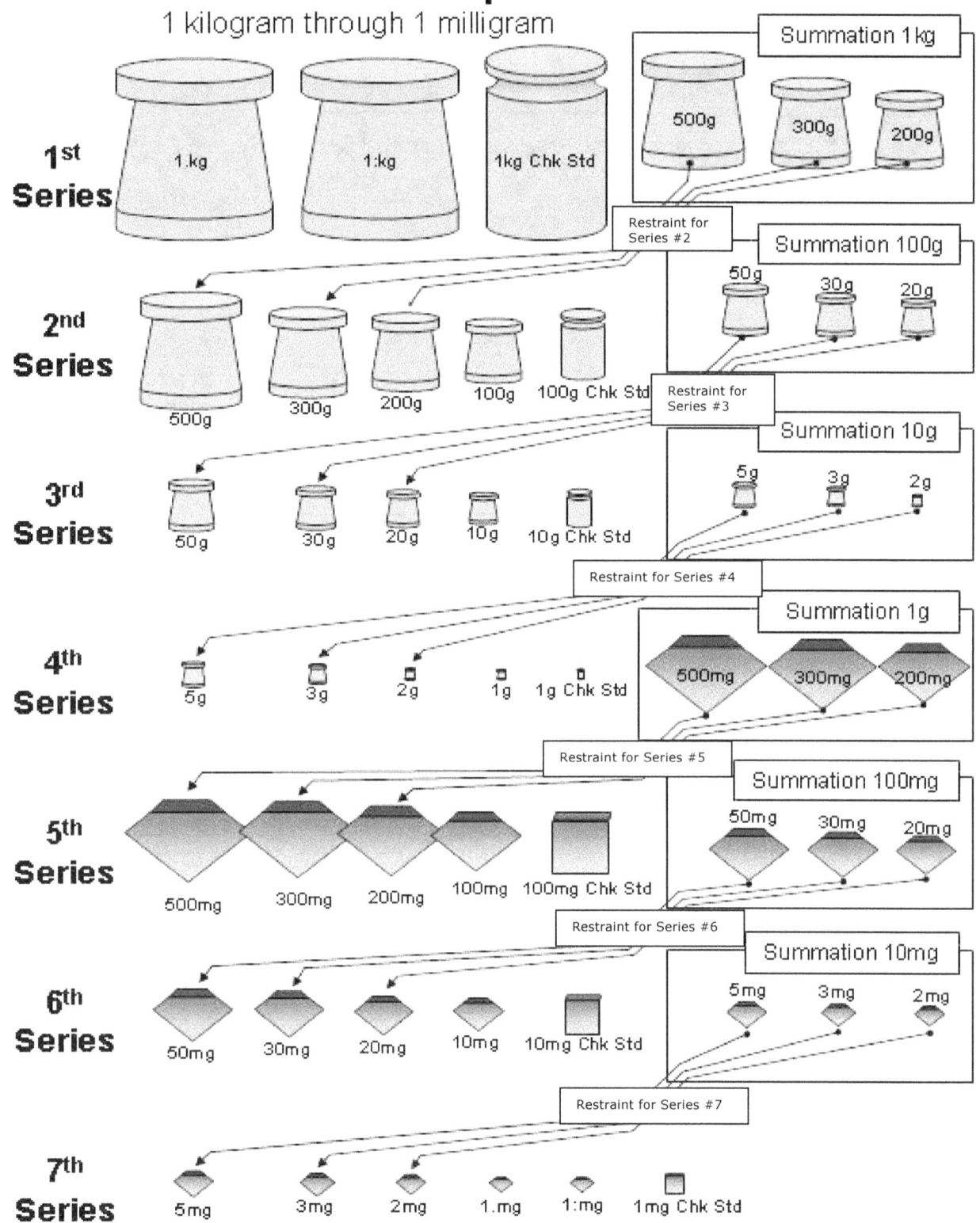

Page 14

March 16, 2005

Measurement Matrix

1 kg through 1 mg

Series 1

	P1.kg	P1:kg	C1kg	Σ1kg	
1st Double Sub	+	−			1 kg load
2nd Double Sub	+		−		1 kg load
3rd Double Sub	+			−	1 kg load
4th Double Sub		+	−		1 kg load
5th Double Sub		+		−	1 kg load
6th Double Sub			+	−	1 kg load

Series 2

	P500g	P300g	P200g	P100g	C100g	Σ100g	
1st Double Sub	+	−	−	+	−		600 g load
2nd Double Sub	+	−	−		+	−	600 g load
3rd Double Sub	+	−	−			+	600 g load
4th Double Sub	+	−	−				500 g load
5th Double Sub	+	−	−	−	−	−	500 g load
6th Double Sub		+	−	+	−	−	400 g load
7th Double Sub		+	−	−	+	−	400 g load
8th Double Sub		+	−	−	−	+	400 g load
9th Double Sub			+	−	−	−	200 g load
10th Double Sub			+	−		−	200 g load
11th Double Sub			+		−	−	200 g load

Series 3

	P50g	P30g	P20g	P10g	C10g	Σ10g	
1st Double Sub	+	−	−	+	−		60 g load
2nd Double Sub	+	−	−		+	−	60 g load
3rd Double Sub	+	−	−	−		+	60 g load
4th Double Sub	+	−	−				50 g load
5th Double Sub	+	−	−	−	−	−	50 g load
6th Double Sub		+	−	+	−	−	40 g load
7th Double Sub		+	−	−	+	−	40 g load
8th Double Sub		+	−	−	−	+	40 g load
9th Double Sub			+	−	−		20 g load
10th Double Sub			+	−		−	20 g load
11th Double Sub			+		−	−	20 g load

Series 4

	P5g	P3g	P2g	P1g	C1g	Σ1g	
1st Double Sub	+	−	−	+	−		6 g load
2nd Double Sub	+	−	−		+	−	6 g load
3rd Double Sub	+	−	−	−		+	6 g load
4th Double Sub	+	−	−				5 g load
5th Double Sub	+	−	−	−	−	−	5 g load
6th Double Sub		+	−	+	−	−	4 g load
7th Double Sub		+	−	−	+	−	4 g load
8th Double Sub		+	−	−	−	+	4 g load
9th Double Sub			+	−	−		2 g load
10th Double Sub			+	−		−	2 g load
11th Double Sub			+		−	−	2 g load

Series 5

	P500mg	P300mg	P200mg	P100mg	C100mg	Σ100mg	
1st DoubleSub	+	−	−	+	−		600mg load
2nd Double Sub	+	−	−		+	−	600mg load
3rd Double Sub	+	−	−	−		+	600mg load
4th Double Sub	+	−	−				500mg load
5th Double Sub	+	−	−	−	−	−	500mg load
6th Double Sub		+	−	+	−	−	400mg load
7th Double Sub		+	−	−	+	−	400mg load
8th Double Sub		+	−	−	−	+	400mg load
9th Double Sub			+	−	−		200mg load
10th Double Sub			+	−		−	200mg load
11th Double Sub			+		−	−	200mg load

Series 6

	P50mg	P30mg	P20mg	P10mg	C10mg	Σ10mg	
1st Double Sub	+	−	−	+	−		60 mg load
2nd Double Sub	+	−	−		+	−	60 mg load
3rd Double Sub	+	−	−	−		+	60 mg load
4th Double Sub	+	−	−				50 mg load
5th Double Sub	+	−	−	−	−	−	50 mg load
6th Double Sub		+	−	+	−	−	40 mg load
7th Double Sub		+	−	−	+	−	40 mg load
8th Double Sub		+	−	−	−	+	40 mg load
9th Double Sub			+	−	−		20 mg load
10th Double Sub			+	−		−	20 mg load
11th Double Sub			+		−	−	20 mg load

Series 7

	P5mg	P3mg	P2mg	P:1mg	P.1mg	C1mg	
1st Double Sub	+	−	−	+	−		6 mg load
2nd Double Sub	+	−	−		+	−	6 mg load
3rd Double Sub	+	−	−	−		+	6 mg load
4th Double Sub	+	−	−				5 mg load
5th Double Sub	+	−	−	−	−	−	5 mg load
6th Double Sub		+	−	+	−	−	4 mg load
7th Double Sub		+	−	−	+	−	4 mg load
8th Double Sub		+	−	−	−	+	4 mg load
9th Double Sub			+	−	−		2 mg load
10th Double Sub			+	−		−	2 mg load
11th Double Sub			+		−	−	2 mg load

Establishing Measurement Controls

Process Evaluation

For each combination of a weighing design, specific load, and specific balance (e.g., a 4-1 design, at a 1-kg load, on an AT 1005 balance), a measurement control process should be defined and data collected to characterize both the standard uncertainty of the process, s_w, and the standard

March 16, 2005

uncertainty for the standard over time, s_t. These values are essential for correctly reducing measurement data and calculating the uncertainty assigned to each mass value in a report. This process is more than simple statistical process control because the assigned values for the standards and check standards are verified for accuracy with each run.

In a surveillance test of reference/working standards from 1-kg to 1-mg, there are seven series as characterized below:

Series	Weighing Design (Tech Note 952)	Decade Load	Balance (examples only)
1	A.1.2 (1, 1, 1, 1) (4-1)	1 kg	AT 1005
1A (option)	A.1.4 (1, 1, 1, 1, 1) (5-1)	1 kg	AT 1005
2	C.2 (5, 3, 2, 1, 1, 1)	600, 500, 400, 200 g	AT 1005
(optional)	A.1.2 or A.1.4	100 g	AT 106, CC111
3	C.2 (5, 3, 2, 1, 1, 1)	60, 50, 40, 20 g	AT 106, CC111
4	C.2 (5, 3, 2, 1, 1, 1)	6, 5, 4, 2 g	UMT-6, CC6
5	C.2 (5, 3, 2, 1, 1, 1)	600, 500, 400, 200 mg	UMT-6, CC6, S5
6	C.2 (5, 3, 2, 1, 1, 1)	60, 50, 40, 20 mg	UMT-6, CC6, S5
7	C.1 (5, 3, 2, 1, 1) or C.2	6, 5, 3, 2, 1 mg	UMT-6, CC6, S5

NOTE: When data is reduced using the mass code, switching balances within a series may result in artificially low process standard deviations and will result in errors in between-time calculations.

As noted earlier, the Advanced LAP problems are the minimum recommendations for collecting data that will begin to characterize the measurement process. Ten complete runs on the reference/working standards, 1-kg through 1-mg (all seven series reduced using the mass code), must be made. An additional series at a 100-g load is essential to establish a secondary starting point for the calibration of 100-g kits. This means that initial data will be collected for at least these eight series.

The section on Calculations in SOP No. 28 shows how to calculate each of the standard uncertainty values that should be entered in the mass code data file once data from the initial ten runs is available. Meanwhile, because statistical data for the new process and check standards may be unavailable, simulated values, based on knowledge of each measurement process, should be used in order for the mass code to reduce data. For the first ten runs, process standard deviation values, based on previous three-ones (also called three in one, abbreviated 3-1) weighing designs or multiple double substitutions for each balance are satisfactory. Using this process, only the size of the

March 16, 2005

uncertainties and the statistical tests will be affected when data is reduced; the mass values are not affected. However, once actual data is available, the simulated data in the data file must be replaced, and final reports generated. All reports should be considered as "draft" during the process of gathering the initial data; the statistics and uncertainty values must be updated prior to performing the final data analysis.

Proper characterization of the measurement process is more critical when using advanced weighing designs in decade series with the mass code than when using routine mass comparisons with one-to-one standards. This is primarily because the statistics used by the mass code distribute the uncertainty from the starting restraint (reference standard) proportionally among all the weights in each series. Also, the standard uncertainty of the process (previously called random error) is distributed among all the weights in each particular series. This type of data reduction allows the mass code to assign smaller uncertainties; however, the validity of these uncertainties is very dependent upon a well-characterized measurement process. A major difference between advanced weighing designs using the mass code, and routine calibrations such as the 3-1 weighing design and the double substitution, is that the mass code uses the uncertainty of the starting restraint only. The 3-1 design and the double substitution use the uncertainty of a single standard at each denomination. In a 1-kg to 1-mg mass code calibration, the starting restraint portion of the uncertainty is distributed among all other denominations and soon becomes negligible around five grams. Therefore, weights below five grams are primarily dependent upon the standard uncertainty of the process when assigning an uncertainty to a test weight.

Data Input

The [1993] Fortran version of the mass code requires input for statistical measurement control parameters in four locations:

1) Line eight: "ran err" or "random error" for the standard is entered first; this hasn't been used and zero must be entered;

2) Line eight: "sys err" or "systematic error" for the standard, s_s, is entered second; this value is taken from the [NIST] calibration report and should be divided by the k-factor that was used in reporting the uncertainty (see calculations in SOP No. 28 for formula to be used when using more than one restraint);

3) Line thirteen: the within-process standard deviation is the same as the standard uncertainty of the process, s_w. This value is used for the F-test and is entered at the beginning of line thirteen on the first series (at the beginning of the line that contains the sensitivity weight data for subsequent series). The value is based on pooled data for observed standard deviations for the process (see calculations in SOP No. 28 for formula to be used); and

4) Line thirteen: the between-time standard deviation, s_b, is entered at the end of line thirteen on the first series (at the end of the line that contains sensitivity weight data for subsequent series) and is the calculated value that measures the variation of the value for the check standard over time, s_t, less the contribution from the standard

March 16, 2005
> uncertainty of the process (see the calculations in SOP No. 28 for formula to be used.)

In Lab Wizard 1.0, the C++, Windows version of the mass code, these four parameters are identified in the data entry area called "Description of Weights." They are identified in Restraint Specifications and Statistical Parameters sections as: "Random error," "Systematic error," "Standard deviation," and "Between std. dev.". These items correspond exactly to the Fortran version of the mass code and are entered in the same location in the data file.

NOTE: Both the Fortran and Lab Wizard 1.0 versions of the mass calibration software have been modified to conform to the NIST policy on uncertainty (see NIST Technical Note 1297) as far as possible. References to "systematic error" have been changed to "type B uncertainty"; references to "random error" have been changed to "type A uncertainty"; and references to "uncertainty" have been changed to "expanded uncertainty." The type B uncertainties are calculated as one standard deviation estimates for systematic error, and the type A uncertainties are calculated as one standard deviation estimates for random error. The expanded uncertainties are calculated as the root sum squares of the type A and type B uncertainties multiplied by a coverage factor of two.

However, to preserve the integrity of the statistical control procedure for mass calibrations, the operation of the code deviates from the NIST policy in the evaluation of type A and type B uncertainties. The policy defines type A (random) uncertainties in a global manner; i.e., as a function of both local phenomena (balance precision and long-term measurement precision) and hierarchical phenomena (uncertainties associated with previously assigned values of reference standards). The test for statistical control for each series requires a standard deviation based only on local phenomena. The code does not distinguish between these two requirements, and it will produce an improper t-test if the hierarchical uncertainties are treated as random components. The solution to this conflict is to distinguish between local and hierarchical uncertainties and to define hierarchical uncertainties as type B uncertainties.[5]

Handling the Output

To establish measurement controls once the mass code has been run, certain data must be extracted from each series and placed in a spreadsheet or database for storage, analysis, and measurement assurance:

> Test Number;
> Operator ID;
> Date of Test;
> Balance ID;
> Restraint ID;
> Check Standard ID;

---

[5] Croarkin, M. C., Internal NIST Correspondence, July & August 1993.

March 16, 2005

    Check Standard Nominal Value;
    Starting Restraint Number;
    Calibration Design ID;
    Average Corrected Temperature in Degrees C;
    Average Corrected Pressure in Pascals;
    Average Corrected Humidity in Percent;
    Average Computed Air Density in mg/cm^3;
    Observed Standard Deviation of the Process;
    Accepted Standard Deviation of the Process;
    Degrees of Freedom;
    F-Ratio;
    Observed Correction of the Check Standard;
    Accepted Mass Correction of the Check Standard;
    t-Value; and
    Expanded Uncertainty assigned to the Check Standard.

When either version of the mass code is run, it produces two files, one called "control" and one called "statis." The "control" file contains a string of data for each series that contains: the month, day, year, check standard identification, observed value for the check standard, balance identification, process standard deviation, degrees of freedom, weighing design identification, average temperature, range of temperature during the measurement, average pressure, range of pressure during the measurement, average humidity, range of humidity during the measurement, air density, range of air density during the measurement, operator identification, and a flag for process control results (0 = ok; a number 1, 2, or 3 (depending on the version of the mass code used) flags that the check standard failed, the observed standard deviation failed, or both failed). This file can be imported to a spreadsheet using a parse function so that each item is entered into a separate cell and saved. The "statis" file contains F-test and t-test data for each series run in the mass code. Unless an assignable blunder is detected in the data, all out of control series must be saved in the control chart file or the statistical limits will gradually become artificially small resulting in an increased number of failed tests.

Each of these files should be saved by another name and/or in another directory immediately after each run of the mass code because the data is not cumulatively saved. These files contain information for the preceding mass code reduction only, and are written over with each run of the mass code.

Reviewing Mass Code Report

Several sections within the mass code report should be reviewed for adequacy. Key areas are observation values, the F-test, and the t-test. If unusually high t-test or F-test values are observed, one should check for data entry errors first.

March 16, 2005

Evaluate the F-test to make sure that the observed standard deviation agrees statistically with the accepted standard deviation of the process. An F-value is quoted and immediately below this value in the report is a statement that shows whether the F-test passed or failed.

Evaluate the t-test to verify that the observed mass correction of the check standard agrees statistically with the accepted mass correction of the check standard. A t-value is quoted and immediately below the t-value in the report is a statement that says whether the check standard is in-control or out-of-control.

Should either test fail and no data-entry errors are found, the series should be rerun. If the process is gradually changing, the t-values or F-values will usually fall in a range from two to nine. If statistical data is graphed properly, trends can be identified before they become critical. If many tests fail, it could suggest a change in the measurement process in which case the data is combined with the other data to define the new measurement parameters. If failed data with no attributable errors is routinely and incorrectly discarded, statistical limits will be artificially reduced. Uncertainty values based on such data will be invalid.

Graphs and Control Charts

Critical Graphs

Standard Control Charts – When advanced mass calibrations are used for surveillance testing, each weight (500 g to 1 mg) involved may be graphed separately. These data and charts can be used to verify calibration values and to determine appropriate calibration intervals. When appropriate for reference/working standards, a new accepted value may be calculated as either the mean of all values, or the predicted value from the linear fit of the data at a time six months in the future.

When the measurement process has been sufficiently characterized and advanced mass calibrations are used for routine calibrations, a graph must be prepared for the check standard at each decade to properly characterize each measurement process. Analysis of the data provides the statistics for calculating the between-time standard deviations and can be used to verify standard and check standard values. The standard deviation over time is calculated from either the standard deviation about the mean, or the residual standard deviation from the linear fit. The standard deviation is later compared with the process deviation to detect if there is a between-time component of error in the measurement process. As noted for surveillance testing, the accepted values can be calculated in one of two ways.

Measurements and control charts for two external or "monitoring" 1-kg check standards are recommended to verify the calibration values for the two 1-kg reference standards. Measurements are made between these two external standards and the reference standards using a 4-1 weighing design. The external kilogram standards may be part of a circulating mass package with recent NIST calibration (as used in proficiency testing) or may be maintained in the laboratory but must receive less frequent (less than 25 % as often) use than the reference standards. Analysis of

March 16, 2005

calibration results over time as recorded on control charts can provide a realistic estimate of calibration uncertainty and allow the investigation of drift for standard values due to time and use. Historical analysis can also help set calibration intervals.

Process charts -- Control charts for the process standard deviation (for each series/balance) are used to establish process variability and an accepted within-process standard deviation, s_w each point can note the degrees of freedom differentiating between series. Standard deviations for each balance are plotted versus time. Plots are critiqued for outliers and degradation. A new accepted value is calculated by pooling the standard deviation for each balance.

Optional Graphs

Summation Graphs -- With a measurement control program in place for 3-1 weighing designs that shows the values for summations of standards, data from the mass code reports may be plotted with those values (when the 3-1 design uses the summation as the check standard). A number of items should be plotted versus time: the mean for the 3-1 summation values, the upper and lower control/warning limits, the mass code summation values and uncertainties, and a line showing the NIST calibration value. This graph may look cluttered, but it provides enormous insight to the relationships between the NIST value, the mass code values, and the 3-1 values for the same group of weights. It also provides a comparison between the separate measurement processes. The 3-1 values may show greater precision than the mass code values. This is because in some cases the 3-1 values may be assigned using a balance with a lower standard deviation of the process than the balance used for the mass code.

F-values -- Graphing the F-values for a particular series can show trends in the process and can evaluate the appropriateness of the assigned process parameters. The F-values are plotted chronologically with the mean. The graph should be analyzed for trends and for uniform distribution above and below the F-value of 1.00.

t-values -- Graphing the t-values of each series allows visualization of the extent of agreement between the observed and accepted value of the check standard. It also permits a comparison with the current measurement process. The t-values are plotted chronologically with the mean. The graph should be analyzed for trends and for uniform distribution above and below the t-value of 1.00.

Proficiency Tests

Interlaboratory comparisons (proficiency tests) should be conducted at least once every four years at the advanced level and may consist of two kits with an entire set of standards: 1 kg through 1 mg. Each regional measurement assurance group will have a schedule that addresses compliance with the WMD PT policy. Charts should be prepared with an "uncertainty bar" format with lines showing the lab average, lab median, the NIST value and NIST upper and lower uncertainties. There should be two charts for each denomination (one for each kit). A third chart should be a scatter plot with a

March 16, 2005

"Youden" analysis, with the center of the circle at the two lab medians and the diameter of the circle based upon the "residual standard deviation" of the participating laboratories. This type of analysis will provide useful information about potential errors in the laboratory, about the uncertainty reported, and about drift of artifacts during the intercomparisons. This information provides opportunity for evaluating a laboratory's capability to meet stated uncertainties.

Evaluation Criteria for Proficiency Tests

Verification of the NIST Value

The NIST value will be verified by comparison with the median of all accepted laboratory results. A standard deviation among laboratories and an overall median are calculated using all participating lab values. Using only those values that fall within the two standard deviation limits, an adjusted standard deviation (a measure of reproducibility between laboratories) and an adjusted round robin median are calculated. These new statistical values are used in the evaluation. Verification of the NIST value (VE) is based on the following formula:

$$(VE) = \left| \frac{(RR_{median} - NIST_{value})}{\sqrt{(1\,sd_{RR}^2 + NIST_{unc(k=1)}^2)}} \right|$$

The normalized error (E_{normal}) concept (which is used internationally) is used to verify the accepted value for each artifact by using the Verification Error (VE). The normalized error is simply a ratio of the difference between the observed and accepted values and the combined uncertainty (at k=1) in the process combined by the root sum square method. The VE value must be less than one to verify the NIST value. When VE is equal to or greater than one, the standards should be calibrated by NIST and the new NIST values and uncertainties should be used to evaluate the proficiency test. The limits on the verification test require the adjusted median value to agree with the reference value within very tight limits. These tight limits are essential for evaluation of proficiency when many laboratories are working at state-of-the art levels.

March 16, 2005
Verification of Laboratory Values

After verification of the NIST value, the following formula is used to evaluate the acceptance of each laboratory value (NVLAP Handbook 150, Procedures and General Requirements, July 2001.):

$$E_{normal} = \left| \frac{(Lab_{value} - NIST_{value})}{\sqrt{(Lab_{unc}^2 + NIST_{unc}^2)}} \right|$$

E_{normal} calculated for a laboratory must be less than one to pass this test. The laboratory uncertainty and the reference value uncertainty are calculated at a 95 % confidence level.

Verification of the Laboratory Precision

This criterion evaluates and validates the reported uncertainty of the laboratory for its suitability to the level against which the laboratory is being evaluated. At the highest level of mass calibration, the uncertainty assigned by the participating laboratory should be less than the tolerance. There are several perspectives regarding the use of tolerance and uncertainty ratios; this is only a general guideline and is not intended to be a requirement. However, if the laboratory will determine compliance to OIML or ASTM standards, the expanded uncertainty at k=2, must be less than 1/3 of the applicable tolerance.

$$U_{lab} < Tolerance \quad OR \quad U_{lab} < Tolerance/3$$

NOTE: Criteria for determining satisfactory/unsatisfactory compliance for proficiency tests may be revised in the future to handle uncertainties based on 95 percent confidence intervals rather than 99.7 percent confidence intervals.

File Management

Mass Code reports that have been generated and printed do not need to be saved on a hard disk; they can take up valuable space on the computer. Instead, a directory may be created to store each data file. Data files take up little disk space and the complete report can be generated as needed.

Once the ten basic runs are made on the standards (LAP Problem 1) and the process parameters are defined, the original data files can be updated with the actual process parameters (replace simulated data with observed data), and final reports can be generated. These new reports will contain the same mass values, but the quoted uncertainties will reflect the true process more closely.

March 16, 2005
Software Management

Distribution

Mass code software will be distributed by the NIST Weights and Measures Division upon acceptable completion of the Advanced Mass Metrology seminar. Lab Wizard software is "personalized" for each laboratory: the name of the laboratory and of the metrologist are embedded in the software and reports are generated containing the name of the laboratory and the metrologist.

Licensing and Software Quality Assurance

Each copy of mass code software distributed by WMD will be serialized, and a list of trained metrologists and their laboratories is maintained. All practical steps have been taken at NIST to ensure correct results when the software is used with proper data files. However, each laboratory must verify this independently and must document the verification. Each licensee must agree to refrain from copying or transferring software to others who have not participated in Advanced Mass training unless WMD gives permission in writing. WMD will only provide technical support to metrologists who have participated in the Advanced Mass training from the NIST Weights and Measures Division.

Updating

The mass code will be periodically updated and new versions will be released to trained metrologists when available.

Approved Weighing Designs

The Weights and Measures Division recognizes a variety of weighing designs such as those found in NBS Technical Notes 844 and 952, and the NIST/SEMATECH e-Handbook. However, weighing designs are used throughout the world with variations from those presented in NIST publications. Metrologists should use good judgment in developing unusual weighing designs and may submit them to NIST for review or validation whenever appropriate. Any designs submitted to NIST for review should be accompanied by sufficient experimental data to provide adequate evaluation. Metrologists should consider, and be able to justify, variations in weighing designs, the selection and use of check standards, length of designs and time restraints particularly with respect to drift, selection of balance, use of sensitivity weights, etc., to suit particular calibration applications. With the use of electronic mass comparators, length of time during a design and fatigue of the operator (which affect design selection) are of less concern than with the older mechanical balances. When developing new designs, another consideration should be that weights of equal nominal values should have the same uncertainty. NIST/WMD strongly recommends designs that incorporate check standards for process and standard verification.

March 16, 2005

Any unusual weighing designs not submitted to NIST for review will be subject to critical review during on-site assessments.

Documentation of Standard Operating Procedures

To help with laboratory compliance to NVLAP Handbook 150, ANSI/NCSL Z540-1-1994, and ISO/IEC 17025, SOP No. 28 "Recommended Standard Operating Procedures for Using Advanced Weighing Designs" was developed and is included with this update as an appendix.

Traceability and Calibration Intervals

Ensuring traceability and providing documentation of how traceability is maintained is a critical concern for customers and for accreditation bodies when evaluating a laboratory. Measurement traceability for mass measurements can be maintained through two reference kilograms that have been calibrated at NIST as long as appropriate mass calibration and measurement assurance procedures are used (and documented). The process described in this document provides guidance for laboratories on how to maintain adequate traceability and uncertainty needs. See GMP 11 and GMP 13 for additional examples of traceability hierarchies and calibration interval requirements. Each laboratory must have an implementation policy that covers the requirements in GMP 11 and GMP 13 to ensure traceability and appropriate calibration intervals. Both of these GMPs are posted on the NIST website.

The assigned LAP problems can be used to initiate an appropriate measurement assurance program and prepare graphs. The data from the analysis and preparation of the graphs must be evaluated against original NIST values for the reference/working mass standards and check standards as appropriate. Additional data collected periodically is added to the original graphs. Data should be updated periodically and evaluated. How often data is updated will depend on the laboratory workload.

Each laboratory must document which standards are used at each level in their traceability hierarchy process, what specific measurement assurance is in place at each step, and how often intercomparisons are conducted. The measurement assurance program, as described, is fully integrated into the actual calibration process. Therefore, this is not an exercise used to provide data for an accreditation body, but actually provides checks on the system that the laboratory can use to ensure that each measurement performed for a customer is accurate and traceable, with validated uncertainties.

It is critical for laboratories to participate in interlaboratory comparisons that provide periodic checks on the measurement process. The data must be correlated with the measurement assurance program to be meaningful. Any discrepancies indicate the need for further investigation and possible need for calibration of the reference standards. Laboratories must participate in this level of intercomparison on a frequency no greater than four years (per WMD policy published in NISTIR

March 16, 2005
7082, Proficiency Test Policy and Plan (for State Weights & Measures Laboratories), G. Harris and J. Gust, January 2004.)

Formulae and Calculations

The following items are calculated using formulae located in SOP No. 28 "Recommended Standard Operating Procedure for Using Advanced Weighing Designs" which is included in this edition as an appendix.

- s_r – The standard uncertainty of the starting restraint in the first series.
- s_w – The within-process standard deviation.
- s_b – The between-time standard deviation for each particular series.
- Effective densities for summation standards.
- Effective cubical coefficients of expansion for summation standards.

SOP 5 is inserted here as Appendix A for printing only; it is kept separate from the document for website downloads.
SOP 28 is inserted here as Appendix B for printing only; it is kept separate from the document for website downloads.

March 16, 2005

SOP 5

Recommended Standard Operations Procedure

for

Using a 3-1 Weighing Design

1. Introduction

   1.1. Purpose

   The 3-1 weighing design is a combination of three double substitution comparisons of three weights of equal nominal value; a standard, an unknown weight, and a second standard called a check standard. (The check standard may be made up of a summation of weights.) The weights are compared using an equal-arm, single-pan mechanical, full electronic, or a combination balance utilizing built-in weights and a digital indication. The specific SOP for the double substitution procedure for each balance is to be followed. The 3-1 weighing design provides two methods of checking the validity of the measurement using an F-test to check the measurement process and a t-test to evaluate the stability of the standard and check standard. Hence, the procedure is especially useful for high accuracy calibrations in which it is critical to assure that the measurements are valid and well documented. This procedure is recommended as a minimum for precision calibration of laboratory working standards that are subsequently used for lower level calibrations and for routine calibration of precision mass standards used for balance calibration. For surveillance of reference and working mass standards and calibration of precision mass standards used to calibrate other mass standards, see SOP 28 for the use of higher level weighing designs.

   1.2 Prerequisites

   1.2.1 Calibrated mass standards, traceable to NIST, with valid calibration certificates must be available with sufficiently small standard uncertainties for the intended level of calibration. Reference standards should only be used to calibrate the next lower level of working standards in the laboratory and should not be used to routinely calibrate customer standards.

   1.2.2 The balance used must be in good operating condition with sufficiently small process standard deviation as verified by F-test values, pooled short term standard deviations, and by a valid control chart for check standards or preliminary experiments to ascertain its performance quality when new balances are put into service.

   1.2.3 The operator must be experienced in precision weighing techniques. The operator must have specific training in SOP 2, SOP 4, SOP 5, SOP 29, and be familiar with the concepts in GMP 10.

March 16, 2005

1.2.4 The laboratory facilities must meet the following minimum conditions to meet the expected uncertainty possible with this procedure:

Table 1. Environmental conditions

Echelon	Temperature	Relative Humidity (percent)
I	20 °C to 23 °C. allowable variation of ± 1 °C Maximum change of 0.5 °C/h	40 to 60 ± 5
II	20 °C to 23 °C. allowable variation of ± 2 °C Maximum change of 1.0 °C/h	40 to 60 ± 10

2 Methodology

2.1 Scope, Precision, Accuracy

This method can be performed on any type of balance using the appropriate double substitution SOP for the particular balance. Because considerable effort is involved, this weighing design is most useful for calibrations of the highest accuracy. The weighing design utilizes three double substitutions to calibrate a single unknown weight. This introduces redundancy into the measurement process and permits two checks on the validity of the measurement; one on accuracy and stability of the standard and the other on process repeatability. A least-squares best fit analysis is done on the measurements to assign a value to the unknown weight. The standard deviation of the process depends upon the resolution of the balance and the care exercised to make the required weighings. The accuracy will depend upon the accuracy and uncertainty of the calibration of the standard weights and the precision of the comparison.

2.2 Summary

A standard weight, S, an unknown weight, X, and a check standard, S_c are intercompared in a specific order using the double substitution procedure. The balance and the weights must be prepared according to the appropriate double substitution SOP for the particular balance being used. Once the balance and weights have been prepared, all readings must be taken from the reading scale of the balance without adjusting the balance or weights in any way. Within a double substitution all weighings are made at regularly spaced time intervals to average out any effects due to instrument drift. Because of the amount of effort required to perform the 3-1 weighing design, the procedure includes the air buoyancy correction.

March 16, 2005

2.3 Apparatus/Equipment Required

2.3.1 Precision analytical balance or mass comparator with sufficient capacity and resolution for the calibrations planned.

2.3.2 Working standard weights and sensitivity weights with valid calibrations traceable to NIST.

2.3.3 Small working standards with valid calibrations traceable to NIST to be used as tare weights.

2.3.4 Uncalibrated weights to be used to adjust the balance to the desired reading range or adequate optical or electronic range for the intended load and range.

2.3.5 Forceps to handle the weights or gloves to be worn if the weights are moved by hand.

2.3.6 Stop watch or other timing device to observe the time of each measurement or the operator is experienced with determining a stable indication. If an electronic balance is used that has a means for indicating a stable reading, the operator may continue to time readings to ensure consistent timing that can minimize errors due to linear drift.

2.3.7 Thermometer accurate to $0.10\,°C$ to determine air temperature.[1]

2.3.8 Barometer accurate to 0.5 mm of mercury (66.5 Pa) to determine air pressure.

2.3.9 Hygrometer accurate to 10 percent to determine relative humidity.

2.4 Procedure

2.4.1 Place the test weight and standards in the balance chamber or near the balance overnight to permit the weights and the balance to attain thermal equilibrium. The equilibration time is particularly important with weights larger than 1 gram. Conduct preliminary measurements to determine the tare weights that may be required, the size of the sensitivity weight required, adjust the balance to the appropriate reading range of the balance indications, and to exercise the balance. Refer to the appropriate double substitution SOP

---

[1] The thermometer, barometer, and hygrometer are used to determine the air density at the time of the measurement. The air density is used to make an air buoyancy correction. The accuracies specified are recommended for high precision calibration. Less accurate equipment can be used with only a small degradation in the overall accuracy of the measurement (See SOP 2).

March 16, 2005

for details.

2.4.2 Weighing Design Matrix

The following table shows the intercomparisons to be made in the 3-1 design, in a matrix format as shown in NBS Technical Note 952, Designs for the Calibration of Standards of Mass, J. M. Cameron, M. C. Croarkin, and R. C. Raybold, 1977.:

Weight ID Comparison	S	X	Sc
a_1 +		-	
a_2 +			-
a_3		+	-
Standard +			
Check Standard			+

This design is represented as design ID "A.1.1" in Technical Note 952, with the exception that the design order is reversed and Restraint B is used. The restraint is another name for the "standard" used in the comparison that may be found in NBS Technical Note 952. This matrix may be useful for anyone using the NIST Mass Code for data reduction. When creating a data file for this design, the design matrix will appear as follows:

```
          restraint 1 0 0
            Check 0 0 1
following series sum      0      0      0
           Report 0 1 1
         1st double sub   1     -1      0
         2nd double sub   1      0     -1
         3rd double sub   0      1     -1
```

2.4.3 Measurement Procedure

Record the pertinent information for the standard, S, unknown, X, and check standard, S_c, as indicated on a suitable data sheet such as the one in the Appendix of this SOP. Record the laboratory ambient temperature,

March 16, 2005

barometric pressure, and relative humidity. Perform the measurements in the order shown in the following table.

Double Substitution	Measurement Number	Weights on Pan	Observation
a_1: S vs X	1	$S + t_s$	O_1
	2	$X + t_x$	O_2
	3	$X + t_x + sw$	O_3
	4	$S + t_s + sw$	O_4
a_2: S vs S_c	5	$S + t_s$	O_5
	6	$S_c + t_{sc}$	O_6
	7	$S_c + t_{sc} + sw$	O_7
	8	$S + t_s + sw$	O_8
a_3: X vs S_c	9	$X + t_x$	O_9
	10	$S_c + t_{sc}$	O_{10}
	11	$S_c + t_{sc} + sw$	O_{11}
	12	$X + t_x + sw$	O_{12}

where:

Variable	Description
t_s	calibrated tare weights carried with S
t_x	calibrated tare weights carried with X
t_{sc}	calibrated tare weights carried with S
sw	calibrated sensitivity weight

3 Calculations

3.1 Calculate the air density, ρ_A, as described in the Appendix to SOP No. 2.

3.2 Calculate the measured differences, a_1, a_2, and a_3, for the weights used in each double substitution using the following formula (note: do not confuse this formula with the calculations used in SOP 4; the signs will be opposite from Option A of SOP 4):

March 16, 2005

$$a_x = \frac{(O_1 - O_2 + O_4 - O_3)}{2} \frac{M_{sw}\left(1 - \frac{\rho_A}{\rho_{sw}}\right)}{O_3 - O_2}$$

where:

Variable	Description
M_{sw}	mass of the sensitivity weight
ρ_{sw}	density of the sensitivity weight

3.3 Calculate the short term within process standard deviation, s_w, for the 3-1 weighing design. This standard deviation has one degree of freedom.

$$s_w = 0.577(a_1 - a_2 + a_3)$$

3.4 Compute the F statistic which compares the short term within process standard deviation, s_w, to the pooled within process standard deviation. (See chapter 8.4 and 8.5 for a discussion of the statistics used in weighing designs.)

$$F\text{-statistic} = \frac{s_w^2}{(\text{Pooled } s_w)^2}$$

The F-statistic so computed must be less than the F-value obtained from an F-table at 99 % confidence level (Table 9.5) to be acceptable. The F-value is obtained from the F-table for numerator degrees of freedom equal one, and denominator degrees of freedom equal to the number of degrees of freedom in the pooled within process standard deviation. If the data fails the F-test and the source of the error cannot be determined conclusively, the measurement must be repeated.

3.5 Compute the observed mass value of the check standard.

Compute the least-squares measured difference d for S_c.

$$d_{S_c} = \frac{-a_1 - 2a_2 - a_3}{3}$$

3.6 Compute the observed mass of S, M_{sc}.

$$M_{sc} = \frac{M_s\left(1 - \frac{\rho_A}{\rho_s}\right) + d_{sc} + M_{ts}\left(1 - \frac{\rho_A}{\rho_{ts}}\right) - M_{tsc}\left(1 - \frac{\rho_A}{\rho_{tsc}}\right)}{\left(1 - \frac{\rho_A}{\rho_{sc}}\right)}$$

3.7 Evaluation of the observed mass of S, M_{sc}.

The mass determined for the check standard should be plotted on the control chart and must lie within the control limits. If it does not, and the source of error cannot be found, the measurement must be repeated. The 'Accepted M_{sc}' is the mean of the historically observed mass values for the check standard.

|Observed M_{sc} – Accepted M_{sc}| > 3sd   Status: Out of Control
2sd < |Observed M_{sc} – Accepted M_{sc}| < 3sd   Status: In Control*Warning
|Observed M_{sc} – Accepted M_{sc}| < 2sd   Status: In Control

Perform an E_{normal} test to compare the mean value of the M_{sc} value from the 3-1 design to a calibration value that has demonstrated measurement traceability for the check standard.

$$E_n = \frac{\left|\overline{M_{sc}} - CalibratedM_{Sc}\right|}{\sqrt{U^2_{M_{sc}} + U^2_{CalibratedM_{sc}}}}$$

The E_{normal} value must be less than one to pass.

3.8 Compute the least-squares measured difference, d for X.

$$d_x = \frac{-2a_1 - a_2 + a_3}{3}$$

3.9 Compute the mass of X, M_x.

$$M_x = \frac{M_s\left(1-\dfrac{\rho_A}{\rho_s}\right)+d_x+M_{ts}\left(1-\dfrac{\rho_A}{\rho_{ts}}\right)-M_{tx}\left(1-\dfrac{\rho_A}{\rho_{tx}}\right)}{\left(1-\dfrac{\rho_A}{\rho_x}\right)}$$

where:

Variable	Description
ρ_A air	density
M_i	mass for weight I
ρ_i	reference density for weight I

3.10 Calculate the conventional mass of X versus the desired reference density of 8.0 g/cm^3 or apparent mass of brass (8.3909 g/cm^3). It is recommended that the conventional mass versus 8.0 g/cm^3 be reported unless otherwise requested. The density of X, ρ_x, must be entered in g/cm^3. (See SOP No. 2)

3.10.1 Conventional mass

$$CM_x \text{ vs. } 8.0 = \frac{M_x\left(1-\dfrac{0.0012}{\rho_x}\right)}{0.999850}$$

3.10.2 Apparent mass versus brass (8.3909 g/cm3 at 20°C)

$$AM_x \text{ vs. } 8.4 = \frac{M_x\left(1-\dfrac{0.0012}{\rho_x}\right)}{\left(1-\dfrac{0.0012}{8.3909}\right)}$$

March 16, 2005

4   Assignment of Uncertainty

The limits of expanded uncertainty, U, include estimates of the standard uncertainty of the mass standards used, u_s, plus the uncertainty of measurement, u_m, at the 95 percent level of confidence. See SOP 29, "Standard Operating Procedures for the Assignment of Uncertainty", for the complete standard operating procedure for calculating the uncertainty. When the 3-1 weighing design is used in conjunction with the Mass Code for data reduction, see SOP 28, "Recommended Standard Operating Procedure for Using Advanced Weighing Designs", for detailed instructions on calculating the uncertainty components which are required by the Mass Code program.

4.1   The standard uncertainty for the standard, u_s, is obtained from the calibration report. The combined standard uncertainty, u_s, is used and not the expanded uncertainty, U, therefore the reported uncertainty for the standard will need to be divided by the coverage factor k. Since only one standard is used as the restraint for the 3-1 weighing design, the uncertainty of the check standard is not included in assigning an uncertainty to the unknown mass.

4.2   Standard deviation of the measurement process from control chart performance (See SOP No. 9.)

The value for s_p is obtained from the control chart data for check standards using 3-1 weighing designs. Statistical control must be verified by the measurement of the check standard in the 3-1 design.

4.3   Other standard uncertainties usually included at this calibration level include uncertainties associated with calculation of air density and standard uncertainties associated with the density of the standards used.

5   Report

5.1   Report results as described in SOP No. 1, Preparation of Test/Calibration Reports.

March 16, 2005

Appendix

3-1 Weighing Design When Tare Weights Are Used
(Densities used to compute air buoyancy correction)
(Air buoyancy correction on the tare weights)

Laboratory data and conditions:

Date	Temperature		
Balance	Pressure		
Load	Relative	Humidity	
Pooled within process s.d., s_w =		Calculated Air Density	
Check standard s.d., s_c =			

Mass standard(s) data:

ID	Mass = N + C (g)	Density (g/cm^3)	Unc$_{(k=1)}$ (mg)	ID	Mass = N + C (g)	Density (g/cm^3)	Unc$_{(k=1)}$ (mg)
N_x				t_x			
M_s				t_s			
M_{sc}				t_{sc}			
sw							

N = Nominal, C = Correction, M = *True Mass*

Laboratory observations:

Balance Observations					
S - X = a_1		S - S$_c$ = a_2		X - S$_c$ = a_3	
S + t_s		S + t_s		X + t_x	
X + t_x S		S_c + t_{sc}		S_c + t_{sc}	
X + t_x + sw		S_c + t_{sc} + sw		S_c + t_{sc} + sw	
S + t_s + sw		S + t_s + sw		X + t_x + sw	
a_1 =		a_2 =		a_3 =	

Note: dotted line represents decimal point.

Calculate "a" values: $$a_x = \frac{(O_1 - O_2 + O_4 - O_3)}{2} \frac{M_{sw}\left(1 - \frac{\rho_A}{\rho_{sw}}\right)}{O_3 - O_2}$$

SOP 5

March 16, 2005

Calculate short term within process standard deviation and conduct F-test:

$$s_w = .577(a_1 - a_2 + a_3)$$

$$F\text{-}statistic = \frac{s_w^2}{(Pooled\ s_w)^2} < value\ F\text{-}table\ 9.5$$

F-test passes? Yes No

Evaluate check standard (by plotting on a control chart or with a t-test):

$$d_{sc} = \frac{-a_1 - 2a_2 - a_3}{3}$$

$$M_{sc} = \frac{M_s\left(1 - \frac{\rho_A}{\rho_s}\right) + d_{sc} + M_{ts}\left(1 - \frac{\rho_A}{\rho_{ts}}\right) - M_{tsc}\left(1 - \frac{\rho_A}{\rho_{tsc}}\right)}{1 - \frac{\rho_A}{\rho_{sc}}}$$

Check standard passes? Yes No

If both F-test and check standard pass the tests, calculate the mass of the unknown test item:

$$d_x = \frac{-2a_1 - a_2 + a_3}{3}$$

$$M_x = \frac{M_s\left(1 - \frac{\rho_A}{\rho_s}\right) + d_x + M_{ts}\left(1 - \frac{\rho_A}{\rho_{ts}}\right) - M_{tx}\left(1 - \frac{\rho_A}{\rho_{tx}}\right)}{1 - \frac{\rho_A}{\rho_x}}$$

$$CM_x\ vs\ \rho_{ref} = \frac{M_x\left(1 - \frac{0.0012}{\rho_x}\right)}{1 - \frac{0.0012}{8.0}}$$

where ρ_{ref} refers to the reference density.

March 16, 2005

Example

3-1 Weighing Design When Tare Weights Are Used
(Densities used to compute air buoyancy correction)
(Air buoyancy correction on the tare weights)

Laboratory data and conditions:

Date	8/18/96	Temperature	21.7
Balance	AT 1005	Pressure	753.5
Load	1 kg	Relative Humidity	45
Pooled within process s.d., s_w=	0.023 mg	Calculated Air Density	1.182 mg/cm^3
Check standard s.d., s_c =	0.10 mg		

Mass standard(s) data:

ID	Mass = N + C	Density (g/cm^3)	Unc$_{(k=1)}$ (mg)	ID	Mass = N + C (g)	Density (g/cm^3)	Unc$_{(k=1)}$ (mg)
N_x	1000 g 7.84 TBD t			x	NA		
M_s	999.99850 g	8	0.0327 mg	t_s	NA		
M_{sc}	1000.0023 g	8	0.0327 mg	t_{sc}	NA		
sw	50.086 mg	8.41	0.0010 mg				

N = Nominal, C = Correction, M = *True Mass*

Laboratory observations:

Balance Observations					
S - X = a_1		S - S$_c$ = a_2		X - S$_c$ = a_3	
S + t_s	10.00	S + t_s	10.30	X + t_x	15.50
X + t_x	15.30	S$_c$ + t_{sc}	14.00	S$_c$ + t_{sc}	14.10
X + t_x + sw	65.30	S$_c$ + t_{sc} + sw	64.10	S$_c$ + t_{sc} + sw	64.00
S + t_s + sw	60.10	S + t_s + sw	60.40	X + t_x + sw	65.60
a_1 = - 5.25829		a_2 = - 3.69845		a_3 = 1.50538	

Note: dotted line represents decimal point.

March 16, 2005

Calculate "a" values:

$$a = \frac{(O_1 - O_2 + O_4 - O_3)}{2} \frac{M_{sw}\left(1 - \frac{\rho_A}{\rho_{sw}}\right)}{O_3 - O_2}$$

Calculate short term within process standard deviation and conduct F-test:

$$s_w = .577(a_1 - a_2 + a_3) = -0.03142$$

$$F\text{-statistic} = \frac{s_w^2}{(\text{Pooled } s_w)^2} < \text{value} \quad F\text{-table } 9.5$$

$$F\text{-statistic} = \frac{-0.03142^2}{0.023^2} = 1.87 < 7.31 \text{value} \quad F\text{-table } 9.5 \text{ d.f.} = 40$$

F-test passes? **Yes** No

Evaluate check standard (by plotting on a control chart or with a t-test):

$$d_{sc} = \frac{-a_1 - 2a_2 - a_3}{3} = \frac{-5.25829 - 2(-3.69845) - 1.50538}{3} = 3.71660 \text{ mg}$$

$$M_{sc} = \frac{M_s\left(1 - \frac{\rho_A}{\rho_s}\right) + d_{sc} + M_{ts}\left(1 - \frac{\rho_A}{\rho_{ts}}\right) - M_{t_{sc}}\left(1 - \frac{\rho_A}{\rho_{t_{sc}}}\right)}{1 - \frac{\rho_A}{\rho_{sc}}}$$

$$M_{sc} = \frac{999.9985 \text{ g}\left(1 - \frac{0.001182}{8}\right) + 0.00371660}{1 - \frac{0.001182}{8}} = 1000.002217 \text{ g}$$

Check standard passes? **Yes** No

If both F-test and check standard pass the tests, calculate the mass of the unknown test item:

$$d_x = \frac{-2a_1 - a_2 + a_3}{3} = \frac{-2(-5.25829) - (-3.69845) + 1.50538}{3} = 5.24014 \text{ mg}$$

$$M_x = \frac{M_s\left(1 - \frac{\rho_A}{\rho_s}\right) + d_x + M_{ts}\left(1 - \frac{\rho_A}{\rho_{ts}}\right) - M_{tx}\left(1 - \frac{\rho_A}{\rho_{tx}}\right)}{1 - \frac{\rho_A}{\rho_x}}$$

$$M_x = \frac{999.9985 \text{ g}\left(1 - \frac{0.001182}{8}\right) + 0.00524014 \text{ g}}{1 - \frac{0.001182}{7.84}} = 1000.006757 \text{ g}$$

$$CM_x = \frac{M_x\left(1 - \frac{0.0012}{\rho_x}\right)}{1 - \frac{0.0012}{\rho_{ref}}} = \frac{1000.006757 \text{ g}\left(1 - \frac{0.0012}{7.84}\right)}{1 - \frac{0.0012}{8}} = 1000.003695 \text{ g}$$

where ρ_{ref} refers to the reference density 8.0 g/cm^3, or conventional mass.

Uncertainty:

$$U = 2 * U_c = 2 * \sqrt{u_s^2 + s_p^2 + u_o^2} = 2 * \sqrt{\left(\frac{0.098 \text{ mg}}{3}\right)^2 + 0.10 \text{ mg}^2 + 0.005 \text{ mg}^2}$$

U = 0.210638 mg = 0.21 mg

C$_x$ = 3.70 mg ± 0.21 mg (Conventional Mass vs 8.0 g/cm^3)

March 16, 2005

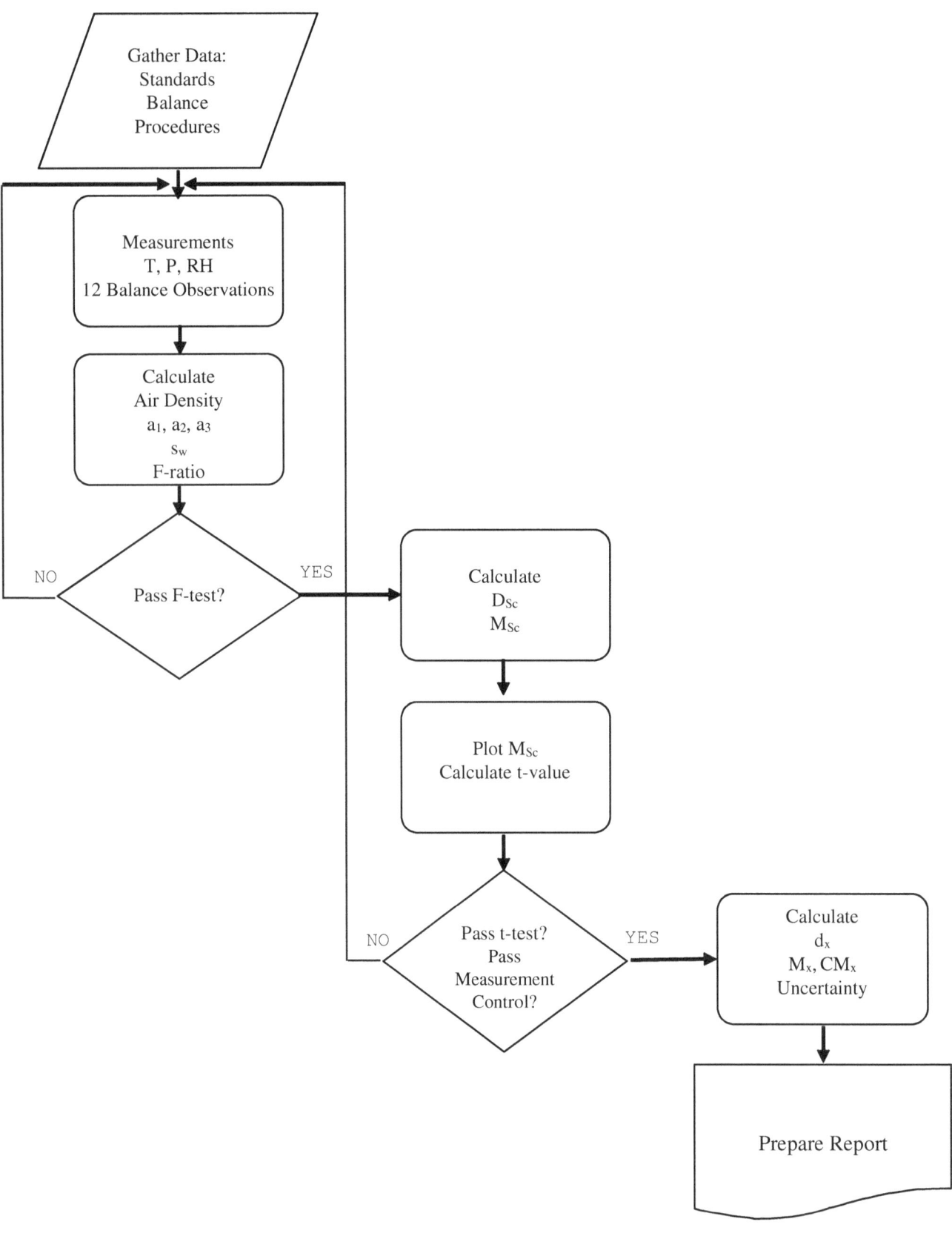

March 16, 2005

SOP No. 28

Recommended Standard Operations Procedure for Using Advanced Weighing Designs

1 Introduction

    1.1 Purpose

        Advanced weighing designs use a combination of double substitution comparisons of weights of equal nominal value or a series of weights in ascending or descending order; standard(s), unknown weights, and an additional standard called a check standard. The weights are intercompared using an equal-arm, single-pan mechanical, full electronic, or a combination balance utilizing built-in weights and a digital indication. The specific SOP for the double substitution procedure for each balance is to be followed. Weighing designs provide two methods of checking the validity of the measurement using an F-test to check the measurement process and a t-test to evaluate the stability of the standard and check standard. Hence, the procedure is especially useful for high accuracy calibrations in which it is critical to assure that the measurements are valid and well documented. This procedure is recommended for precision calibration of laboratory working standards that are subsequently used for lower level calibrations, for routine calibration of precision mass standards used for calibration of other mass standards, and for surveillance of mass reference and working standards.

    1.2   Prerequisites

        1.2.1   Verify that valid calibration certificates are available for the standards used as restraints in the test.

        1.2.2   Verify that the standards to be used have sufficiently small standard uncertainties for the intended level of calibration. Reference standards should only be used to calibrate the highest level of working standards in the laboratory and should not be used to routinely calibrate customer standards.

        1.2.3   Verify that the balance used is in good operating condition with sufficiently small process standard deviation as verified by F-test values, pooled short term standard deviations, and by a valid control chart for check standards, or preliminary experiments to ascertain its performance quality when new balances are put into service. See NISTIR 5672[1] for a discussion on the performance levels expected for use of these procedures

---

[1] Fraley, K. L., Harris, Georgia G. L., NIST IR 5672, Advanced Mass Calibration and Measurement Assurance Program for State Calibration Laboratories, March 2005.

as part of a laboratory measurement assurance program to ensure traceability of laboratory standards.

1.2.4 Verify that the operator is experienced in precision weighing techniques, and has had specific training in SOP 2, SOP 4, SOP 5, SOP 29, and is familiar with the concepts in GMP 10. Further, the operator must have been trained in the creation of data files and the operation of the NIST Mass Code when it is used for data reduction as recommended. Example data sets and sample observation sheets are available in the Advanced Mass Seminar offered by the NIST Weights and Measures Division.

1.2.5 Verify that the laboratory facilities meet the following minimum conditions to meet the expected uncertainty possible with this procedure:

Table 1. Environmental conditions

Echelon	Temperature	Relative Humidity (percent)
I	20 °C to 23 °C. allowable variation of ± 1 °C maximum change of 0.5 °C/h	40 to 60 ± 5

2 Methodology

2.1 Scope, Precision, Accuracy

This method can be performed on any type of balance using the appropriate double substitution SOP for the particular balance. Because considerable effort is involved, this weighing design is most useful for calibrations of the highest accuracy. The weighing design utilizes a combination of double substitutions to calibrate a single unknown weight, or a group of related weights in a decade. This method introduces redundancy into the measurement process and permits two checks on the validity of the measurement; one on accuracy and stability of the standard and the other on process repeatability. A least-squares best fit analysis is done on the measurements to assign a value to the unknown weights. The standard deviation of the process depends upon the resolution of the balance and the care exercised to make the required weighings. The accuracy will depend upon the accuracy and uncertainty of the calibration of the restraint weights and the precision of the comparison.

2.2 Summary

A restraint weight, S, in some cases two restraint weights, S_1 and S_2, an unknown weight, X, or group of unknown weights, and a check standard, S_c are compared in a specific order typically using the double substitution procedure although other procedures may be appropriate. The balance and the weights must be prepared according to the appropriate double substitution SOP for the particular balance being used. Once the balance and weights have been prepared, all readings must be taken from the reading scale of the balance without adjusting the balance or weights in any way. Within a double substitution all weighings are made at regularly spaced time intervals to minimize effects due to instrument drift. Because of the amount of effort required to perform weighing designs, the procedure includes an air buoyancy correction using the average air density as determined immediately before and after the weighings, drift-free equation for calculating the observed differences, correction for the cubical coefficient of expansion when measurements are not made at 20 °C, an average sensitivity for the balance over the range of measurements made, and the international formula for air density.[2]

2.3 Apparatus/Equipment Required

2.3.1 Precision analytical balance or mass comparator with sufficient capacity and resolution for the calibrations planned.

2.3.2 Reference standard weights (usually starting at 1 kg or 100 g), calibrated check standards for each decade (e.g., 1 kg, 100 g, 10 g, 1 g, 100 mg, 10 mg, 1 mg for the seven series between 1 kg and 1 mg), working standard weights and sensitivity weights with valid calibrations traceable to NIST.

2.3.3 Small standard working standards with valid calibrations traceable to NIST to be used as tare weights. Note: The calculations performed by the mass code do not take into consideration the value of any tare weights used in the weighing design. Additional calculations will be required when tare weights are used.

2.3.4 Uncalibrated weights to be used to adjust the balance to the desired reading range or adequate optical or electronic range for the intended load and range.

2.3.5 Forceps to handle the weights or gloves to be worn if the weights are moved by hand.

2.3.6 Stop watch or other timing device to observe the time of each measurement or the operator is experienced with determining a stable indication. If an electronic balance is used that has a means for indicating a stable reading, the operator may continue to time readings to ensure consistent timing that can minimize errors due to linear drift.

2.3.7 Thermometer accurate to 0.10 °C with recent calibration certificate traceable to NIST to determine air temperature.[3]

---

[2] Formula for the Density of Moist Air, (CIPM-81/91). This equation is published in SOP 2. The difference between Option A and Option B in SOP 2 is less than the uncertainty associated with air density equations.

[3] The thermometer, barometer, and hygrometer are used to determine the air density at the time of the measurement. The air density is used to make an air buoyancy correction. The accuracies specified are recommended for high precision calibration. Less accurate equipment can be used with only a small degradation in the overall accuracy of the measurement (See SOP 2).

March 16, 2005

- 2.3.8 Barometer accurate to 0.5 mm of mercury (66.5 Pa) with recent calibration certificate traceable to NIST to determine air pressure.

- 2.3.9 Hygrometer accurate to 10 percent with recent calibration certificate traceable to NIST to determine relative humidity.

- 2.3.10 Computer with sufficient processing capability and memory.

2.4 Procedure

- 2.4.1 Place the test weight and standards in the balance chamber or near the balance overnight to permit the weights and the balance to attain thermal equilibrium, or use a thermal soaking plate next to the balance with weights covered. Thermal equilibration time is particularly important with weights larger than 1 gram. An alternative heat source such as a heat lamp may further improve temperature stability in front of the balance. Conduct preliminary measurements to determine the size of the sensitivity weight and any tare weights that are required, adjust the balance to the appropriate reading range of the balance indications, and to exercise the balance. Refer to the appropriate double substitution SOP for details.

- 2.4.2 Weighing Designs

    The table below shows the most common comparisons to be made as referenced in NBS Technical Note 952, Designs for the Calibration of Standards of Mass, J. M. Cameron, M. C. Croarkin, and R. C. Raybold, 1977. Each series is characterized by the number of observations, n, the degrees of freedom, *d.f.* associated with the standard deviation, the number of weights in each design, k (not shown in this table), the number of restraints (standards), and check standards, along with appropriate positions within the design*.

    *Positions for check standards must be carefully considered as subsequent equations may be dependent on the position of use.

SOP 28

March 16, 2005

Table 2. Common weighing designs

Design ID	Description	n Observations	d.f. Degrees of freedom	Restraint Position	Check Std Position*
A.1.1	3-1 Weighing Design[4]	3	1	1	3
A.1.2	4-1 Weighing Design	6	3	1, 2	3 or 4
A.1.4	5-1 Weighing Design	10	6	1, 2	3, 4, or 5
A.2.1	6-1 Weighing Design	8	3	1, 2	3, 4, 5, or 6
C.1**	5, 3, 2, 1, 1 Design (descending)	8	4	1, 2, 3	5 or 4
C.1	5, 3, 2, 1, 1 Design (ascending)	8	4	5	4
C.2	5, 3, 2, 1, 1, 1 Design (descending)	11	6	1, 2, 3	4, 5, or 6
C.2	5, 3, 2, 1, 1, 1 Design (ascending)	11	6	6	4 or 5
C.9**	5, 2, 2, 1, 1 Design (descending)	8	4	1, 2, 3, 4	5
C.9	5, 2, 2, 1, 1 Design (ascending)	8	4	5	4
C.10	5, 2, 2, 1, 1, 1 Design (descending)	8	3	1, 2, 3, 4	5 or 6
C.10	5, 2, 2, 1, 1, 1 Design (ascending)	8	3	6	4 or 5

**If these designs are NOT the last in a series, there is no position for a check standard.

The "restraint" is another name for the standard used in the comparison. Matrices are shown in Technical Note 952. Determine the best design prior to beginning the series. The series shown allow calibration of any commonly found set of mass standards in either the 5, 2, 2, 1 combination or the 5, 3, 2, 1 combination.

2.4.3 Measurement Procedure

Record the pertinent information for all weights being intercompared on a suitable data sheet unless an automated data collection system is being used to collect the data and create a data file. Record or collect the laboratory ambient temperature, barometric pressure, and relative humidity immediately before and immediately after each series of intercomparisons.

3 Calculations

Calculations are completed by the NIST Mass Code as described in NBS Technical Note 1127, National Bureau of Standards Mass Calibration Computer Software, R. N. Varner, and R. C. Raybold, July 1980, with updates to conform to the international formula for calculating air density and the ISO Guide to the Expression of Uncertainties, 1993, and

---

[4]Design a.1.1. with inverted order (y3, y2, and y1), with restraint in position 1 (B) is detailed in SOP 5.

minor error corrections to the original code. The code is the same as that used by the NIST Mass Group for routine calibrations. The code performs two statistical tests (t-test and F-test) to verify both the value of the restraints and check standards, and to verify that the measurement process was in control during the comparisons.

3.1 Calculating Effective Densities and Coefficients of Expansion for Summations[5]:

Some designs use a summation mass and sometimes the individual masses of this summation will be constructed from different materials that have different densities and coefficients of expansion. The following equations will be used to calculate the effective density and effective coefficient of expansions for the summation that will be needed as input for the data file. The subscripts 5, 3, and 2 refer to the individual masses that comprise the summation. This approach may also be needed with a 5, 2, 2, 1 combination.

$$\textit{Effective Density} = \frac{M_5 + M_3 + M_2}{\left(\frac{M_5}{\rho_5}\right) + \left(\frac{M_3}{\rho_3}\right) + \left(\frac{M_2}{\rho_2}\right)}$$

$$\textit{Effective Cubical Coefficient of Expansion} = \frac{\left(\frac{M_5}{\rho_5}\alpha_5\right) + \left(\frac{M_3}{\rho_3}\alpha_3\right) + \left(\frac{M_2}{\rho_2}\alpha_2\right)}{\left(\frac{M_5}{\rho_5}\right) + \left(\frac{M_3}{\rho_3}\right) + \left(\frac{M_2}{\rho_2}\right)}$$

Table 3. Variables for equations above

Variable	Description
M	Mass (g)
ρ	Density (g/cm^3)
α	Cubical Coefficient of Expansion (/$^\circ$C)

4 Assignment of Uncertainty

The NIST Mass Code generates uncertainties as a part of the data reduction. Proper input in the data file is critical for obtaining valid results and is dependent upon a well characterized measurement process. See NIST IR 5672 for a discussion on the input for standard uncertainties in the data file.

4.1 Calculating the standard uncertainty, u_s, of the starting restraint in the first series:

Usually the starting restraint will be one or several 1 kg (or 100 g) mass standards that have NIST calibrations and density determinations. The uncertainty of the

---

[5] Jaeger, K B., and R. S. Davis, NIST Special Publication 700-1, A Primer for Mass Metrology, November 1984.

standard as stated on a calibration report is divided by two or three, dependent on the confidence interval stated in the calibration report.

One starting restraint scheme (a single starting standard), where U_s is the uncertainty from NIST which must be divided by the proper coverage factor, k.

$$u_s = \frac{U_s}{k_{factor}}$$

Multiple starting restraint scheme with standards calibrated at the same time against the same starting standards, i.e., dependent calibration (more than one starting standard):

$$u_s = \frac{U_{s1}}{k_{factor1}} + \frac{U_{s2}}{k_{factor2}},$$

or

$$u_s = \frac{U_{s1}}{k_{factor1}} + \frac{U_{s2}}{k_{factor2}} + \frac{U_{s3}}{k_{factor3}}, \; etc.$$

Multiple starting restraint scheme with standards *NOT* calibrated at the same time as the starting standards, i.e., independent calibration (more than one starting standard):

$$u_s = \sqrt{\left(\frac{U_{s1}}{k_{factor1}}\right)^2 + \left(\frac{U_{s2}}{k_{factor2}}\right)^2},$$

or

$$u_s = \sqrt{\left(\frac{U_{s1}}{k_{factor1}}\right)^2 + \left(\frac{U_{s2}}{k_{factor2}}\right)^2 + \left(\frac{U_{s3}}{k_{factor3}}\right)^2}, \; etc.$$

March 16, 2005

4.2 Calculating the within-process standard deviation, s_w, for a particular series:

For each particular weighing design, the observed within process standard deviation, s_w, along with its degrees of freedom, d.f., is pooled using the technique described in NIST Handbook 145 section 8.4.

$$s_w = \sqrt{\frac{df_1 \, s_1^2 + df_2 \, s_2^2 + ... + df_k \, s_k^2}{df_1 + df_2 + ... + df_k}}$$

4.3 Calculating the between-time standard deviation for each particular series (s_b):

Establish a standard deviation (s_t) for each check standard over time. If a plot of the check standard shows no apparent drift, the between-time standard deviation may be calculated. The following formulae are used to calculate the between-time standard deviation for the particular series. If s_b^2 is less than zero, then s_b equals zero.

4.3.1 For the 3-1 design with a single restraint, and a check standard that is either another single weight or a summation, the between time standard deviation is calculated using the following formula. The check standard may be in any position.

$$s_b = \frac{1}{K_2}\sqrt{s_t^2 - K_1^2 s_w^2}$$

$$K_1 = 0.8165$$
$$K_2 = 1.4142$$

$$s_b = \frac{1}{1.4142}\sqrt{s_t^2 - 0.8165^2 s_w^2}$$

4.3.2 Using a 4-1 design with two restraints, and the check standard is the difference between the two restraints, the next equation may be used to calculate the between-time standard deviation. If another weight in the series is used as the check standard, another equation is needed.

$$s_b = \frac{1}{K_2}\sqrt{s_t^2 - K_1^2 s_w^2}$$

$$K_1 = 0.7071$$

$$K_2 = 1.4141$$

$$s_b = \frac{1}{1.4141}\sqrt{s_t^2 - 0.7071^2 s_w^2}$$

4.4.3 Using a 4-1 design with two restraints, and with a single check standard occupying any of the remaining positions, the next equation may be used to calculate the between-time standard deviation.

$$s_b = \frac{1}{K_2}\sqrt{s_t^2 - K_1^2 s_w^2}$$

$$K_1 = 0.6124$$

$$K_2 = 1.2247$$

$$s_b = \frac{1}{1.2247}\sqrt{s_t^2 - 0.6124^2 s_w^2}$$

4.3.4 Using a 5-1 design with two restraints, and the check standard is the difference between the two restraints, the next equation may be used to calculate the between-time standard deviation. If another weight in the series is used as the check standard, another equation is needed.

$$s_b = \frac{1}{K_2}\sqrt{s_t^2 - K_1^2 s_w^2}$$

$$K_1 = 0.6325$$

$$K_2 = 1.4142$$

$$s_b = \frac{1}{1.4142}\sqrt{s_t^2 - 0.6325^2 s_w^2}$$

4.3.5 Using a 5-1 design with two restraints, and with a single check standard occupying any of the remaining positions, the next equation may be used to calculate the between-time standard deviation.

$$s_b = \frac{1}{K_2}\sqrt{s_t^2 - K_1^2 s_w^2}$$

$$K_1 = 0.5477$$

$$K_2 = 1.2247$$

$$s_b = \frac{1}{1.2247}\sqrt{s_t^2 - 0.5477^2 s_w^2}$$

4.3.6 In the second series (C.2), six weights are involved (500 g, 300 g, 200 g, 100 g, Check 100 g, and a summation 100 g). Calculate the standard deviations of the mass values for the Check 100 g (s_t) and plot the results to evaluate the presence or lack of drift. If no drift is present, the following formula is used to calculate the between-time standard deviation for this series and all subsequent C.2 series. Subsequent series include the following check standards: 100 g, 10 g, 1 g, 100 mg, 10 mg, 1 mg. If s_b^2 is less than zero, then s_b equals zero.

$$s_b = \frac{1}{K_2}\sqrt{s_t^2 - K_1^2 s_w^2}$$

$$K_1 = 0.3551$$
$$K_2 = 1.0149$$

$$s_b = \frac{1}{1.0149}\sqrt{s_t^2 - 0.3551^2 s_w^2}$$

4.3.7 If a C.1 series is used, the following equation is used to calculate the between-time standard deviation when the check standard is in either of the last two positions:

$$s_b = \frac{1}{K_2}\sqrt{s_t^2 - K_1^2 s_w^2}$$

$$K_1 = 0.4253$$
$$K_2 = 1.0149$$

$$s_b = \frac{1}{1.0149}\sqrt{s_t^2 - 0.4253^2 s_w^2}$$

4.3.8 The between-time formulae shown here are those that are most common and are for descending series only. If another restraint or check standard is used, or if an ascending series is used, another formula will be needed. These formulae are statistically derived, based on the least squares analysis of the weighing design, and assume a normal, non-drifting distribution of measurement results. Equations for some other weighing designs may be calculated using the NIST Electronic Engineering Statistics Handbook. Section 2.3.3.2 "Solutions to Calibration Designs" gives an overview for deriving the solutions to weighing designs. It also provides the unifying equation for s_b (it is called s_{days} in the electronic handbook). To clarify the difference in terminology and notation the unifying equation for s_b is presented as:

$$s_{days} = \frac{1}{K_2}\sqrt{s_2^2 - K_1^2 s_1^2} \qquad \begin{array}{c} s_{days} \equiv s_b \\ s_1 \equiv s_w \\ s_2 \equiv s_t \end{array} \qquad s_b = \frac{1}{K_2}\sqrt{s_t^2 - K_1^2 s_w^2}$$

March 16, 2005

Section 2.3.4.1 "Mass Weights" provides the solutions for 17 weighing designs used for decreasing weight sets, 6 weighing designs for increasing weight sets and 1 design for pound weights. K_1 is located in the portion of the solution titled "Factors for Repeatability Standard Deviations", and K_2 is located in the portion titled "Factors for Between-Day Standard Deviations".

5   Report

Report results as printed in Tables I and II as generated by the Mass Code. Actual text of the mass code report must be modified for each laboratory in order to be ISO/IEC 17025 compliant.

www.ingramcontent.com/pod-product-compliance
Lightning Source LLC
Chambersburg PA
CBHW081741170526
45167CB00009B/3897